Layman's Universe

By Smithstonyen

Smithstonyen

Text copyright © 2014 Kenneth Smith

All Rights Reserved

Dedication

To Any Who Take Action to Achieve a Better Outcome for the Human Race In Part or As a Whole
And
To All Who "Stay the Course" and at Least Consider that Written

Preface

The Author

Smithstonyen is a miners son and virgin writer (or first timer and like they say "I was just testing it out, then this happened" Got a cradle handy?) who tries to give the reader in this instance, his own "Layman's Viewpoint" on The Universe, leaving the reader to ascertain whether he has "lost it altogether" or if it really is "out there" with him. You may be thinking "What's this all about and why does he think that he knows better than all the scientific gobble de gook that can be mustered on this subject"? Well the latter is the reason for this being written because he could not accept most, if not all of the scenarios put forward culminating in the statement that the "Big Bang Theory" meant that all we survey came from "Nothing from

Nowhere in No Time" is how he would put it. This said, some of his concepts to explain the realities of our existence will no doubt hit a bum note with some if not all of any readers, especially those of the scientific community but this is an attempt to stimulate debate into a more logical approach to this subject.

Dare he suggest that reading "Life around Black Holes" would give more of an insight into him as a being living on a Planet within said Universe and contemplating "Why we as a species **perpetrate or allow** that which continually surrounds us to the detriment of our Humanity".

Table of Contents

Chapter (1)

THE UNIVERSE; as I see it, from a layman's viewpoint

Chapter (2)

LIGHT as I see it

Chapter (3)

BLACK HOLES a matter of fact?

Chapter (4)

TIME is of the essence

Chapter (5)

FORCES within the Universe

Chapter (6)

RELIGIAN AND THE PARANORMAL within the Universe

Chapter (7)

LIFE with many questions

Chapter (8)

FOOD CHAIN goblin up; DARK ENERGY; DARK MATTER and DARK FLOW

Chapter (9)

ELECTRICITY and a conclusion

Chapter (10)

UNIVERSE SCENARIOS the Big Bang Theory?

Chapter (11)

THE EXPANDING UNIVERSE or is it?

Chapter (12)

TIME TRAVEL a vision of?

Epilogue

LAYMAN'S UNIVERSE

Chapter (1)

THE UNIVERSE; as I see it, from a layman's viewpoint

Some time ago, I decided to put my thoughts on the following topic into writing and owing to the preceding book being in the main "My Two Pennyworth" on life as I seesaw it, perhaps I shall be forgiven for expanding my remit, so to speak by writing this section on the ultimate expansion, well that's the theory anyway of others. Perhaps the truth is I have really lost it but I'm sure it's "Out There" somewhere!

Well! Here goes, buckle up and don't hold your breath, It's Blast Off!!!

Let us now consider the limitations of our human attributes

In general, we only really believe that which our senses allow us to; that is, seeing, hearing, touching, smelling and tasting being the five most recognizable.

More recently, we have all had to accept, more and more bizarre realities, such as radio, television, mobile phones and microwaves, together with all the derivatives of electricity, light manipulations and various other radiations.

The end product of all such activity is a medium, which our senses can perceive, such as the audio, pictorial or derived heat of a process.

You may be thinking; what has this got to do with the universe?

I think that to appreciate our own limitations is fundamental to our potential development and the acceptance of things beyond our present understanding. There is a commonly known saying that, necessity is the mother of invention and I would argue along with others, that the crude reverse also applies, if you don't use it; you lose it. (We hear you coming from far and wide but we've played a blinder by living in caves, we must be "Bats" is their next cry but such is life; is just one illustration that we all know).

Evolution fits well into these two statements

The basics of evolution could, in the past, have been seen as a natural self preservation process but I would propose that the natural element of this is now superseded by our present ability, **to manipulate our own physical attributes,** should we desire, or be allowed to do so (Although of course, this could be the natural process of evolution in itself).

I believe that we have to consider the human race as a unique entity upon this planet, being what appears to be the only life form with desires beyond our immediate survival and the capability to radically affect the natural processes.

Several aspects of the universe need to be considered; let's begin with:-

WHAT IS IT?

Having considered this over many years, I believe that the most feasible explanation, if we accept, the big-bang theory, is that the physical properties, which are apparent to our senses, could only be answered by the acceptance of these factors, namely:-

1). NOTHING cannot exist

2). A POSITIVE SOMETHING can exist

3). A NEGATIVE SOMETHING can exist

It is a big step but if we accept these factors as possibilities, the answer to some of the anomalies of the universe could be addressed.

We are all aware of the forces unleashed by what in "my day" was called the splitting of the atom. Now consider this to be as, a gnats cough in a global storm, when related to the forces unleashed by the rending apart of opposites on the scale of the big bang.

This rending apart would have, by its extremely violent nature, scattered the opposites far and wide and released an immense amount of energy in all its possible forms.

These opposites I will refer to as positive and negative in an attempt to clarify my meaning of one cancelling the other on their combination in equating units.

As in the **manmade descriptive tool** of mathematics:-

Positive, 1 + Negative, 1 = Naught, 0

In this particular Reference:-

Positive equates to MATTER (The material UNIVERSE)

Plus + equates to COMBINING ENERGY (The GLUE of the fabric)

Negative equates to ANTI-MATTER (Non material UNIVERSE)

Naught equates to THE FABRIC OF THE UNIVERSE (Space?)

MATTER being all single or combinations of the known or yet unknown elements of the universe

ENERGY being, all forms of energy that, presently, exists in the material universe

ANTI-MATTER being, that which, when combined with matter becomes the fabric of the universe

THE FABRIC OF THE UNIVERSE being, that which, exists throughout the whole of the universe

Author's (that's me) Note; although I have used the term "Fabric" to describe space as it is more frequently described; I feel it may be necessary to say that I do not "see" this (nor do you by the way) as a linear medium to which the term is frequently attached in the material universe but more as a totally encompassing function of the whole "Universe" or any number of these.

Chapter (2)

LIGHT as I see it

I believe that light is the passage of energy, at a frequency of vibration, for want of a better description, within the fabric of the universe. Where no elements of the material universe are present in the form known to us, the speed of this passage is referred to as, the speed of light (In a vacuum) and my interpretation of a vacuum being **a space which holds no element of the material universe.**

I will also refer to the speed of light as being the speed limit of the fabric of the universe. Owing to our senses, being able to readily detect light, this is the radiated energy form to which we relate more easily. Other forms of radiation, some not routinely detected by our natural senses, consist of energy transported within the fabric of the universe. These radiations conform to the same speed limits set by light and their speed within the fabric of the universe is also, that known or referred to as "The speed of light".

Owing to the speed reductions of radiated energies, being relatively minor within some elements (such as light through glass and radio waves through air), also that their speed limit remains the same in the material world, a conclusion that the fabric of the universe extends throughout the material universe and that it therefore exists within us all, is not excluded.

Other forms of energy transport, which are limited to the material universe, that do not exist within the fabric of the universe alone, such as sound, heat conduction and convection, which all rely on a material presence, have different speed parameters. The speed of sound, for instance, being limited by the vibration characteristics of the material medium through which it travels, is a minute fraction of the speed of light. Note that heat does travel through space by these methods but only as radiation at the speed of light due to the absence of materials for other forms of transport as I am confident you are aware, I am not trying "to tell a grandfather (or grandmother) how to suck eggs" because I am one; a grandfather not an egg or egg head come to think, shame about that.

Chapter (3)

BLACK HOLES a matter of fact?

My definition being that these are **pools** (for want of a better word) of anti-matter which in turn can be defined as the exact opposite of matter.

I note that scientists have recently intimated that these may exist within all galaxies, which I have long thought to be the case.

My belief is that they can be explained by the interpretation:-

Black holes are the negative opposite of the material universe that was "created" at the occurrence of the big bang. The opposites, matter and anti-matter, becoming separated by vast distances, owing to the release of the phenomenal energy, **which had**

combined them and on release, acting to push them apart.

Effectively one immense tear in the fabric of the universe becoming a multitude of smaller tares (black holes or anti-matter pools), spread out over the universe.

Over time the energy of the big bang was dissipated by the vast distances involved, equilibrium, of sorts, being reached, centered on each black hole. These consisting of sufficient matter and energy to recombine with the black hole anti-matter and thereby forming GALAXIES made up of a truly amazing array of assemblies and combinations of elements plus energy forms, all under the influence of the forces emanating from the central black hole, as we know them today.

This model of the universe allows the consideration that the force exerted by the black hole is actually the force I will refer to as the attraction of opposites, rather than the commonly acknowledged gravity, which requires the black hole to have a truly **unbelievable** density and not to have any physically detectable presence other than by absence of physical entities.

The analogy, which I think best represents a black hole, together with its actions is that of a **weaver / spinner of the fabric of the universe** (space spinner) or more brutishly galaxy gobbler.

Absence of light and other emissions, from a black hole, in this model need not be by virtue of immense gravity but by the following possibilities dare I intimate probability:-

1. Absence of the medium in which they travel. The reason being, the possibility that the; black hole displaces the fabric of the universe (with anti matter?) and as such the black hole is the

only place where the fabric of the universe is not present.

2. All sources of energy are totally absorbed by its actions. Owing to the nature of the beast, all energy sources are fully used up in the process of combining matter and anti-matter. (There are none left to emit). Any observed ejections of matter or energy, which appear to emanate from the vicinity of a black hole, could be the result of excesses from the recombining (spinning) processes, this being inevitably absorbed at some future time of requirement as these progresses.

Another major spin off related to this model of the universe is that, it may answer why the black hole would diminish rather than expand as it absorbed the galaxy.

By creating that which it exists within (fabric of the universe)

From that existing within it (matter and energy of the galaxy)

Together with that of which it is constituted (anti-matter)

THE BLACK HOLE PERFORMS THE ULTIMATE VANISHING ACT

Perhaps, one of the most disturbing aspects of this perceived situation is that not only the black hole vanishes but also the galaxy and all that it contained. This includes us together with all that we hold dear.

I would like to proffer the following question for your deliberation:-

Is it conceivable that the human race has been selected in order to redress this situation, and in so doing, maintain existence of the universe, as we know it rather than reaching a point of status quo at some future time, when all galaxies have been absorbed into the fabric of the universe? The resultant of the latter situation being that change would cease to happen and therefore time, would cease to exist; along with us of course whoops! Pass the Scotties or should it be toilet role?

I am aware, that at this present time experiments within the scientific community are actively pursuing a search for THE GOD PARTICAL whatever this is!

I would particularly advise great caution in this field of activity, considering the resultant situation of creating a tear in the fabric of the universe. This being the possibility of creating their own BIG BANG, which could far exceed any, previously encountered on this earth.

I HOPE THEY KNOW WHAT THEY'RE DOING!

Don't YOU!!!

Or OUR TIME HAS COME may soon become a reality or at least is fast approaching see what I mean about toilet roll now!

There are reports that a potential Black Hole is thought to exist, at the position where, what is said to have been, the most violent event known to have occurred, since the Big Bang. This would effectively indicate an explosion large enough to create the conditions to **Tear the Fabric of the Universe** and in so doing could actually be described as a miniature Big Bang. I surmise, the severity of the explosion being insufficient to disperse the anti matter, leaving a Black Hole, with influence over the matter around it. Seems like the birth of a Galaxy being described, to me!

Chapter (4)

TIME is of the essence

What is time?

My immediate thoughts on this subject are:-

Time is change

The previous reference to time, seizing to exist, requires further investigation. Time not being a physical entity cannot in itself exist, it being merely a means of description relating to changes occurring. Human's main preoccupation with time today, usually relates to reversing it in an attempt to pre-empt the lottery draw but on a more serious note it is actually a regulator, which governs our lives as we all know.

There are many references made to the passage of time commonly used to describe our perceptions of this such as:-

Time does fly when you are enjoying yourself.

A watched pot never boils.

Where? Has the time gone?

Etc

Now let's consider several scenarios:-

Waiting for something to happen - little change therefore time drags.

Examples of this are dental and doctors waiting rooms, where an attempt to alleviate this by the introduction of audio, visual or play themes. These are in effect, increasing changes to accelerate time perception.

More recently, the use of computers within our daily lives, has given the phenomena of the apparent acceleration of time, which I believe to be afforded by our concentration on the myriad of changes occurring to our visual inputs; or the Switch it on, and your time is gone syndrome.

Boredom is another obvious example of time, dragging due to a lack of change. The sales propaganda statement of when it's gone it's gone also fits the profile of time. The implication being that time is a one-way street (you cannot turn it back to suit a particular circumstance).

There are many references to looking back in time; even to the event which created the universe but this is only looking back at that which has already happened; not a reversal of time. In reality, this is a delay of the information reaching our senses, being interpreted by them as present events.

When all matter, and anti-matter has recombined, consuming all the energy to do so, a status of no change will in effect, stop time; because the concept of forever is not conceivable then, time will begin with, the next big bang, tearing apart the fabric of the universe. And so on, and so on! (Well that's one interpretation of present knowledge), but not necessarily mine.

Chapter (5)

FORCES within the Universe

Gravity

This is defined to be the force of attraction acting between all physical bodies and is related to the mass of each body; its effects can be scientifically measured and observed.

The reason for its existence is a lot harder for us to appreciate (even though it enables us to exist in our present form) and in our present habitat.

My model of the universe would represent this force, gravity, as being-natures preservation strategy in an attempt to prolong the existence of the material universe by pulling material objects together, to increase mass and strengthen the resistance to the force which is destroying them, in their present form i.e. Their nemesis "Black Hole".

Today, we can observe the effect of what we describe as black holes these being interpreted scientifically as having an astronomical gravitational force, and therefore that its mass is also proportionately so. The name Black hole being derived from, I assume; the absence of light- Black and the ability to hold material Hole.

An alternative suggestion could be that the force acting from the black hole is not gravity but the attraction of opposites, that is positive to negative or in this scenario, matter to antimatter.

Let us name this the dark force. Considering that it is in the process of obliterating our material universe, by drawing it into the proximity of its counterpart, in order that the combination process can be achieved. As previously described, this entails the use of all the energy within the material and indicates that the phenomena of gravity, preventing light escaping is not relevant, light being one form of energy used up in the combination process and therefore not emanating from it.

In brief, could it be that gravity is natures force to gather materials together in order to gain mass to oppose the dark force trying to obliterate it.

Nature

My definition of Nature is the force acting to preserve the material universe.

Perhaps entity would be a better description than force.

Life

Could it be one of nature's products or tools to aid its preservation strategy?

Human Life form

At present, the selected result of life, to carry forward nature's strategies; we can only hope that this is a wise choice and we don't make a botch of it in the future if we haven't already.

Energy

Energy is the force that enables the fabric of the universe to exist; it can take many forms and thereby perform many tasks.

Light

Light is one form of energy, some portions of which are made aware to us by our senses; it is a very important part of a range of radiated energy forms, which appear to travel through the fabric of the universe without any appreciable loss, thus enabling energy transfer within the material universe; this in turn enables the existence of life, as we now know it.

Radiant heat

This is a form of energy transport, acting within the fabric of the universe, enabling energy transfer between materials over vast distances; it also forms part of the spectrum of light being in one non-visible part of this spectrum usually referred to as infrared.

Conduction and convection

These are the heat transport systems within the elements and their combinations of the material universe.

You would be forgiven for thinking that this is basic physics, taught at early school lessons but if we now consider that, radiations of both heat and light (the more apparent ones) occur within our material world, with losses attributed to its materials then logically; this would seem to indicate, that the fabric of the universe exists throughout all materials as previously hypothesized. That is, the fabric of the universe (space) does not end where our atmosphere begins but exists throughout the material universe, and probably, beyond the farthest reaches of it.

Size (Does matter)

Many words are used to describe size, within the parameters of our existence, but as with all descriptions of quantity or distance, a relationship between us or some other known factor is considered in this description; for example, a light year certainly makes mountains into molehills.

I would propose the following questions:-

How small is the limit at which existence ends?

Can Infinity exist; in the terms of distance or any other concept?

Can forever exist, in terms of time?

However, seriously, I may consider these questions; I certainly don't profess to have any answers to them but what I can indicate is that there appears to be nothing that would prevent the fabric of the universe containing other systems of galaxies, formed by other big bangs and so on and so on ad Infinitum or not; as the case may be. I could also contemplate the concept that atomic structure could be the solar system to some other entity. But then again, that's me!

Chapter (6)

RELIGIAN AND THE PARANORMAL
within the Universe

My view is that, from a practical viewpoint, religion (least said soonest mended) is a tool, devised by humans in an attempt to guide us forward by giving rules to follow, which are in turn **supposed** to progress civilization: (now look how that turned out!). For example; the most basic rule, probably being: - We don't eat each other. Take note, even this took quite a time to filter down to some.

This said there appear to be some similarities between the features of the universe and those associated with religion; take good and evil, for instance, these could be represented by the material universe filled with light as the good, with the black hole gobbling it up, as the evil dark forces, allowing no light to escape.

Heaven and Hell also fit well into this scenario, with GOD portrayed as Nature, which is attempting to design us in particular, with a view to reverse the demise of the material universe and maintain the Heavens whilst the sinister all consuming Black Holes being portrayed as Hell!

I would also propose the possibility that various phenomena, described by many under such headings as:-

Paranormal activity
Déjà vu
Foreboding
Ghosts
Psychic powers etc. etc.

Could be some anomaly within the fabric of the universe (if this exists throughout the whole of the material universe of which we are all a part) and which some small minority may be more susceptible to its influences; so no more witch hunts please!

The only "stakes" we should have interest in, are those on the plates with the chips, peas and onions or perhaps more to the point (no don't carve this on it) the lady ones i.e. Miss stakes; get the gist or is it jest of this! And there's plenty of those to find without "Cluso" being enlisted.

Chapter (7)

LIFE with many questions

Life poses many more fundamental questions such as:-

What is it?

What does it relate to?

How did it begin within an inorganic environment?

Why did it evolve into the many forms we have on our planet?

Was plant life, initiated by nature, to sustain the later animal life?

Does animal life, become inevitable once plant life is established?

Are we, the ultimate biological life form?

Will we evolve to decide our own destiny?

Is everything that exists, part of one gigantic food chain?

What is Life?

I would suggest that in the context of biology, life is the ability to recreate itself and in so doing, independently create a future differing from the past; the essence of the resulting changes, being to effectively create time, to which life can relate its activities.

This is more than the ability to replicate itself, in accordance with a DNA blueprint; it is the ability to redesign its self, according to a perceived requirement of its senses, both conscious and unconscious. OR IS IT?

My conclusion, from the preceding, is that a force or sense, of which they have no conscious knowledge, exists within all life forms, however primitive or complex. This force gives the appearance of having the ability to make decisions and act upon them by re-programming DNA, without the direct involvement of the life form concerned, no matter how primitive or complex this may be.

An answer to the predominant questions (in my mind anyway) such as those, relating to plant and insect symbiotic relationships for example, where a plant appears to have designed itself to suit the purposes of a particular insect. Or is it that an insect has designed itself to suit that particular plant?

You may now be thinking that this is just describing evolution, as we now know it but I believe there is the possibility that this model of life, could involve an influence from within the fabric of the universe, if this fabric exists throughout the material universe, and therefore within every life form, in that material universe.

We could call this influence nature, or even God.

I mentioned reprogramming deliberately, to illustrate a similarity, which I do not believe to be coincidental, between the present positions of our life form, with relation to nature. In evolutionary terms, recently, we have developed technology exponentially, as some of the secrets of the universe have become apparent.

This has led to the development of the computer, which is, to all intents and purposes, a replica of our life forms attributes or at least our mental ones. No doubt the future will see the development of a more viable facsimile of the human form to aid us, or otherwise, with mundane mechanical tasks initially but finally do a takeover bid, as portrayed in science fiction.

Leaving this latter point aside, computers have enabled us to develop technology to such an extent that we can now mimic nature or God, as you wish, with only our reasoning powers to limit us. We can reprogram DNA to design a new life form or alter our own and, given the chance, no doubt we will, probably using "space" exploration as the excuse or even more likely as part of the defense budgetary proposals of some country or other.

The question left to decide is:-

Has nature designed us to do this or have we taken over?

Chapter (8)

FOOD CHAIN goblin up DARK ENERGY; DARK MATTER and DARK FLOW

To develop the food chain question; we all are aware of the food chains within our own planets environment. These can be, in general, related to plant and animal life of many different species, existing in many differing environments and they are, in fact, a range of the known elements EATING EACH OTHER, to sustain their particular interests.

There are parallels to this within the universe as we know it, possibly the ultimate food chain being the one which begins with the Big Bang, as the birth of the universe and ends with the top predator namely The Black Holes and these in turn becoming extinct as the material universe is devoured. ONLY TIME WILL TELL

Dark Energy

This latter "Food Chain" resulting in the expansion of the Universe as previously described could perhaps be a more rational explanation for the many theories (I will call them that for now) which abound with regard to the presence of what is generally termed "Dark Energy". This expansion of The Fabric of The Universe could be interpreted as it being repulsed from the vicinity of the particular "Black Hole" creating it for instance and in turn the "Spinning Process" would inevitably increase over time as the food supply increased due to the gradual but again inevitable decrease of its distance, owing to the gravitational and/or the proposed "Attraction of Opposites" forces, within the galaxy, to the "Devouring Entity"; thus accounting for the apparent increase in the rate of expansion over time of our

Universe. In brief(s) "much like my stomach as I get older, now there's a picture to contemplate, or not of cause.

    Dark Energy as an intrinsic part of the makeup of "The Fabric of The Universe" would also be detected at a constant density, irrespective of The Universes expansion, because the expansion in this model is due to the amount of increase in the quantity of this fabric being generated. In short(s) this time I would propose that IF The Universe is considered, as I write this, to consist of "Its Fabric" which is constituted from Matter and Antimatter with Dark Energy as its Glue or even just a constituent part; then this would explain why the scientific processes have derived the probability that "Dark Energy" mass (I will term it) far outstrips that of the known "Material Universe". Like they say "You do the maths" but I can't be bothered; well that's my excuse!

To simplify:-

If the whole of the Fabric of The Universe exists throughout all of it and it consists of "Matter + Energy + Antimatter" where "Dark Energy" is one of its components then logic decrees that owing to the vast amount of this fabric that already is in evidence and therefore Dark Energy also whose mass will far exceed the mass of our known Material Universe; the latter will inevitably reduce as the black holes gobble up the Materials in the form which we can detect and effectively spread them, probably uniformly, throughout space as they join the existing components of the Fabric of The Universe: or summat like that!

Dark Matter

It is said that scientists have noted that the rotational forces acting on galaxies do not fit into the basic "Laws of Gravity" owing to the outer materials travelling at the same speed as the inner materials, which contradicts the expectation that outer matter would travel at a reduced speed. In order to address this anomaly, "Dark Matter" has been sort of invented, so to speak, such that the known laws can still be rationalized. Now consider my proposed Universe and let's assume the force described as "The Attraction of Opposites" does exist between all the matter of the galaxy and the anti-matter of the Black Hole at its centre, then relate this force to the spinning motion of the Black Hole perimeter such that it acts in a manner which would effectively drag the matter it would encounter to a higher or lower velocity as determined by its normally expected gravitational rotation and in so doing explain the apparent deviation from the "Laws of Gravity".

Before dismissing the previous hypothesis ask yourself which has more credence; this or "Dark Matter" which has to be **in all the right places** and is a figment of the imagination to manipulate scientific results which do not comply with the physical laws determined by themselves. At this stage of the game there can be no proof of either and I am not a scientist so perhaps you should go with the dark side as it's not worth warring over stars; (Unless it's a trilogy).

Dark Flow

More recently Dark Flow has entered the arena, so to speak and some may think that this may detract from my proposals for an alternative Universe but on consideration of what this flow consists of i.e. apparent excessive movement of groups of galaxies in random locations within "Our Universe"; my simplistic but practical viewpoint leads me to the possibility that this could be just an expansion of "The Black Hole in The Galaxy Centre" theory. To spell this out, imagine a very much larger "Black Hole" then could this not have influence over galaxies rather than stars and in so doing cause the movement of galaxy clusters as it becomes the contender for the title of "Top Predator" in the "Cosmic Food Chain" or are there even larger holes in this theory Black or otherwise? Perhaps you prefer to believe other universes are worm holing or summat!

The real question to debate is "Does it Matter?" and to my way of thinking, wouldn't their energy be better spent creating say "A cure for some unmentionable disease" at this stage of our evolution but then again that's me. There is one point which I think Confucius may have made:-

Better to communicate in a language that at least most can understand; then more of us will do just that!

Chapter (9)

ELECTRICITY and a conclusion

What is it?

Basically electricity, as I see it, (well not actually see it!) is a transport of energy.

How fast does it travel?

In vacuum electricity is recognized as having speed equal to that of light in a vacuum.

This speed can be reduced by interaction with the physical universe, much as is the case when considering light. My logic is that this speed limit being the same for light, radio waves, other radiations and electricity, even when being considered within the material universe, would indicate the limiting factor to be present within materials. That is the fabric of the universe is everywhere, with the possible exception of a black hole, which could be conceived as being a hole in the fabric of the universe.

In conclusion:-

This is a layman's hypothetical analysis of the known factors concerning The Universe or is it just A Universe.

It predicts that each galaxy will have a black hole, which equates to the amount of anti-matter to absorb all the material and energy within its region of influence (The Galaxy) and in so doing the Galaxy and Black Hole will cease to exist in that form and become part of the fabric of the universe.

At the point in time, when all the physical and energy entities are absorbed the repair of the fabric of the universe will have effectively used up all the anti-matter, therefore all black holes will have ceased to exist and a status quo situation would occur.

Also at this point because there would be no perceivable difference between the present and the future, TIME would cease to be i.e. no change, no time; until of course, the next big bang and so on and so on ad Infinitum.

Heaven only knows how long this would continue.

Forever is such a long time but I am confident someone will try to find a number to evaluate it and others will follow in its wake, my thoughts on this are probably best left unsaid.

Chapter (10)

UNIVERSE SCENARIOS the Big Bang Theory?

Let us now consider one of the scientific scenarios put forward as a serious contender for the description of how the universe came to be, normally referred to as:

The Big Bang Theory

I believe that a normal rational human brain (I hope that I have one, then perhaps not) at our present state of development or evolution, would not accept the possibility that the whole of the universe was created instantaneously from an infinitely small, infinitely dense (what I don't know) in the middle of nowhere. A lot of something from nothing in zero time is certainly NOT a concept that I would seriously consider and I think that I would, to say the least, not stand alone in this viewpoint.

To propose a mathematical description of this occurrence is equally bizarre to me, so much for the singularity idea! THE MIND BOGGLES well, mine does anyway. Doesn't Yours?

A far more rational explanation for this particular theory being conceived in the first instance is that, if the "right people" propose a scenario, then others will follow and try to make reality fit, irrespective of the distortions necessary, perhaps these are the space distortions some refer to.

To illustrate this take the real world situations relating to, what are termed Modern Art and Impressionism, again I must stress that this is obviously, only my viewpoint but it appears to me, that here is an amazing example of the concept, "they say it's Wunderbar SO IT MUST BE"!

Again as illustration, it is a fact that an impressionist daub, titled "The Scream" was sold for a staggering $ 119,922,500 on May 2nd 2012 and then you are informed that it is only "one of a set of four". Well that's all right then! The artist in question I assume is dead, serious artists must fear for their life, as the motive builds. It is my belief that had a school child produced "The Scream" it would probably have ended "in the bin" and a session with a psychiatrist being advised as the next logical step.

It must appear to some that I am describing a confidence trick in relation to the above purchase but the trick is to decide who is conning who, there are many factors which come to mind, with regard to why the purchase was made, was it:-

To gain Kudos in their particular sphere of activity?

To give the proverbial "finger" to others who cannot afford a crust?

Just because they could?

A form of investment? (The seller must be laughing all the way to the bank).

OR

Was it because the buyer wanted to sit staring at it to the end of their days, in order to prolong the appearance of time passing slowly, in that depressive state and effectively extend their lifespan? (I don't think so either).

This latter point added to indicate, how diverse and ridiculous a motive for an action might be.

Personally, I rather favor the financial motive, with the optional thought that; if I am daft enough to do this, then there must be someone dafter to perpetuate the situation! This has the drawback of the possibility of being left "holding the hot potato" and with "egg on ones face" but at least one wouldn't be starving! One thing I have learnt, as the years roll bye (No! I am not going to buy a "masterpiece" to alleviate this, nor could I of course) is that to trust blindly to the acknowledged "expert" is NOT the way forward. In my experience, I could quote many examples where this trust is brought into question and in fact can be shown to be totally wrong in some cases, which can lead to quite serious consequences. THE GLOBAL FINANCIAL SYSTEM IS PROBABLY THE BEST ILLUSTRATION, AT THIS TIME, OF TRUSTING THE EXPERTS. See what I mean!

BUT, I DIGRESS:-

Suffice it to state, there is instant fame & fortune, for the person who actually achieves, GETTING SOMETHING FROM NOTHING!

Smithstonyen

(FROM not FOR as I've left the financial section) if you get my drift; as the snowman said to the hedgerow.

Chapter (11)

THE EXPANDING UNIVERSE or is it?

In my world of skepticism even this scenario requires closer examination and personally I would have to be convinced that ALL other options had been eliminated before this conclusion was decided to be factual.

Owing to the vast distances involved and the relative time element in making these measurements, I just wonder if the option of the illusion of expansion could in fact be attributed to a contracting universe.

I will now attempt to open up the debate on this point, if only for my own curiosity. The following illustration is an attempt to do this:-

First consider galaxies – These appear to act as individual units of mass, because of the forces acting within them to bind their constituent parts. Taking this as fact and that earth is anchored within one of them; my logic indicates (to me anyway) that distances however determined are referenced to our galaxy in some way.

Now consider The Big Bang Theory it seems fair to assume that the forces unleashed at this perceived event would be actively expanding the known universe, at least up to the point in time when these forces were dissipated by distance to the extent that the force we call gravity, which acts supposedly between all masses throughout the universe, overcomes the diminishing expanding force and effectively reverses the expansion.

What I think would be the natural assumptions are that, the forces released at the inception of the Big Bang would act in all directions and thereby distribute the elements accordingly, in so doing this would create a situation where gravity would act towards the "point" of origin, in the final analysis. Given this situation, the galaxies would be drawn towards the "point" of origin, their relative speeds being determined by their location and mass, i.e. speed would increase as distance or mass decreased. Also as the galaxies moved towards the "point" of origin, there would be a myriad of gravitational forces acting between them, again according to location and mass but for the sake of clarity in the scenario which follows, this can be taken as of no final consequence but could obviously lead to some erroneous conclusions in relation to an expanding or contracting universe.

Now using a simplistic viewpoint let us consider three particular galaxies together with the point of origin of the Big Bang, these being roughly aligned and located in the following positions:-

Galaxy (A) in the outer reaches of the universe

Galaxy (B) in the middle region of the universe (say our galaxy)

Galaxy (C) in the inner region of the universe

Point of origin (O) in the centre of the universe (or thereabouts)

In an expanding universe:-

Distance between (A) and (B) is increasing

Distance between (B) and (C) is increasing

Distance between (C) and (O) is increasing but very difficult to ascertain, (O) has No Physical Properties

In a contracting universe:-

Distance between (A) and (B) is increasing. Due to relative velocity

Distance between (B) and (C) is increasing. Due to relative velocity

Distance between (C) and (O) is decreasing but very difficult to ascertain

It can readily be seen that when we consider the origin having "No Physical Properties", this particular scenario could very easily lead to a wrong conclusion as to whether the universe was contracting or expanding. This even more evident when viewed in relation to our lifespan and even recorded data available.

Logically it would seem that if the relative distances were obtained between Galaxies and say, Earth (as a reference point) in all directions, there may be a small arc in which this would indicate a reduction in this distance with a contracting universe. To add to any confusion this assumes a point of origin whereas this could instead be an extremely vast tear in the fabric of the universe, which could greatly affect convergence with gravitational contraction. Throw into this mix the inter-galactic gravitational forces which "must" abound within the universe and a definitive answer becomes much less likely in relation to: Expanding universe?

What I consider to be the most feasible explanation, should we accept that the Universe is expanding fits logically into my previous Black Hole Scenario, were this is seen as a "Space Spinner" regenerating the Fabric of the Universe from its original constituents. If we now accept the possibility that this "Fabric" does not have elastic properties and requires a finite amount of "space" (pardon the pun), could **we stretch our imagination instead** to believe or at least consider the possibility, that with Black

Holes abounding throughout the Universe, spinning away at their task, the fabric increase could conceivably be sufficient to achieve "An Expanding Universe", or at least the appearance of one!

My logic sees this as the gradual return to a "Universe" of dimensions equating to those before the "Big Bang"; if indeed there was a single one!

Well that's my "Layman's Best Guess Theory"!

Since I started putting these thoughts to paper, there has been an announcement of a breakthrough in the science of particle physics relating to evidence of the existence of a much sort after particle, namely The Higgs Boson, which was predicted many years previously by Higgs, on a theoretical basis, in reference to what I will term The Blueprint Of The Universe.

Let me suggest from the outset, that this in no way impinges on the possible scenarios put forward in this journal, much rather the reverse, being my interpretation of this scientific exploration, when considered in relation to; what the discovery is and what it took to make the observation. Namely:-

Hundreds of top scientists over a considerable period

A huge Particle Accelerator (In terms of earthly machines)

A massive power requirement

All this, just to observe the particles aftermath rather than the actual particle, or so we are led to believe

In fact I would propose that this particle and others are just part of the processes taking place within a Black Hole to combine matter and antimatter into The Fabric of the Universe which permeates the entire known universe. The similarity of the rim of a Black Hole to a Particle Accelerator has not gone unnoticed! (By me anyway)

Please, someone advise them "not to attempt to build one that size" at least not before all on earth have clean water to drink!

Confucius says: - (Or was it me?)

Better to change the theory to fit the facts, than the facts to fit the theory!

Perhaps I should have taken more interest in my "School Sciences Lessons"! I am quietly confident many will think so BUT at least they will have read my book! Like you, if you got this far, forever an optimist isn't I.

You Decide!

Chapter (12)

TIME TRAVEL a vision of?

More recently I was privy to a program which focused on the physics relating to the speed of light and within this program there was reference to its absolute value being a cornerstone of scientific thought in relation to the UNIVERSE, moreover the scientific data was discussed with regard to the experimental results which "apparently" defied the Laws of Physics by indicating that; The Speed of Light could be surpassed. Whereas I don't profess to have the wherewithal to dispute these so called "Laws" I was left scratching my head over some of the reasoning and mathematical gobbledegook proposed by some of the scientific community as explanation for the phenomena of neutrinos travelling beyond this magical velocity.

Let's examine these neutrinos; whoops easier said than done, the little blighters are whipping in and out of other dimensions according to some, creating havoc no doubt or is there a lot of this (doubt) as the theorists are let loose to come forth with time travel in the form of getting there before you set off; perhaps this would solve our transport difficulties, "at a stroke"; my response is "beam me up" and let's get on with it! Or perhaps it's time to fetch **Ida Ont Tinkso out** once more.

To examine the phenomena of "looking back in time" a little closer (He should have gone to the opticians) is my first thought on this subject, as you may by now have guessed but in order not to be "seen" as treating this subject frivolously I will attempt to give clarity to my "layman's viewpoint", on earth as it is in heaven so to speak. If an action has taken place and at some time later you perceive this action then logic indicates that your speed of travel to the source of the action cannot reduce the time from the action taking place plus the time taken to perceive it; therefore YOU CANNOT TRAVEL BACK IN TIME and that's a FACT. Obviously "looking" does not equate to "travelling" and as such requires a lot more consideration; whereas I may be persuaded to accept that a star we see in the night sky could no longer be in existence owing to the time taken for the light, via which we perceive its existence, to reach us and this seems fair enough. But looking back to the origins of the Universe or "Big Bang", if indeed there was one, definitely does not fit into that scenario; well in my book anyway; after all the background "Cosmic Radiation" could just be us detecting "The Fabric of The Universe's" own inherent output(s). Couldn't it!

THE END
OR IS IT JUST ANOTHER
BIG INNING!

Epilogue

The question uppermost in my mind with regard to this is "What have I just written, is it fact or fiction; real or false; kiddology or comedy of a like, you're having a laugh variety"? I haven't got a clue and in saying that we could possibly come to some agreement after all but I sincerely hope that in some way I have brought some things you may not have dwelled on in the past into a little better focus or perhaps a little adjustment of lens position is all that's required.

Perhaps I could expand on this in the future with my logical approach being stretched to the "Outer Limits" (hopefully not of your endurance) as I could move even closer to "Science Fiction" by attempting the prediction of the way it will be for those who follow in our "Wake" so to speak. Who knows? Certainly not me!

Bye for now anyway!

And don't forget what to do before or after cleaning your teeth!

Up's! That's in the other book "Life around Black Holes" sorry!

Smithstonyen